Classroom Supplement to
REGRESSION ANALYSIS AND ITS APPLICATION

A Data-Oriented Approach

RICHARD F. GUNST
Department of Statistics
Southern Methodist University
Dallas, Texas

ROBERT L. MASON
Energy Systems Research Division
Southwest Research Institute
San Antonio, Texas

TSUSHUNG A. HUA
Department of Mathematical Sciences
Northern Illinois University
DeKalb, Illinois

MARCEL DEKKER, INC.   New York and Basel

Library of Congress Cataloging in Publication Data

Gunst, Richard F. [date]
    Classroom supplement to Regression analysis and its application.
      1. Regression analysis. I. Mason, Robert Lee [date]. II. Hua, Tsushung A. [date].
III. Gunst, Richard F. [date]. Regression analysis and its application. IV. Title.
QA278.2.G85 Suppl.   519.5'36   81-12599
ISBN 0-8247-1694-9   AACR2

MARCEL DEKKER, INC.

270 Madison Avenue, New York, New York 10016

Current printing (last digit):

10 9 8 7 6 5 4 3 2 1

PRINTED IN THE UNITED STATES OF AMERICA

PREFACE

One of the major objectives of *Regression Analysis and Its Application: A Data-Oriented Approach (RAA)* is to liken a comprehensive regression analysis to the conduct of scientific research. By treating a regression analysis similar to a research study, the oversimplified view of regression as a series of mechanical computations is seen to be inadequate. A thorough regression analysis typically involves problem formulation and data collection, model specification and verification of assumptions, and estimation and variable selection. Each of these topics is developed in *RAA* by outlining or deriving theoretical principals underlying the techniques utilized and also by illustrating their application on a variety of data sets. This supplement offers readers the opportunity to gain further familiarity with and understanding of the major concepts and procedures advocated in the text.

The supplement consists of two major components, each one keyed to the chapters in *RAA*. First, additional exercises are provided which reinforce or expand on topics covered in the individual chapters. These exercises treat both applied and theoretical topics and usually can be selected without reference to one another in order to gain greater expertise with any of the individual techniques of interest to the reader. Many of the applied exercises parallel examples in the text but use different data sets.

The second major component of the supplement is intended to focus on a comprehensive analysis of the two data sets in Appendix B of *RAA*. Data Set B.1, the Solid Waste Data, will be analyzed through a series of assignments which are geared to specific sections of the various chapters. Solutions or comments on the analyses will be provided in Appendix B of this supplement. Although similar questions and suggested analyses will be posed for Data Set B.2, the Mortality and Pollution Data, no solutions or conclusions will be given. The reader will soon discover that the analyses of these data sets are neither unique nor straightforward. Successful completion of the projects will introduce the reader to a variety of concerns which one faces in most regression analyses, the successful resolution of which rarely seem to be unique or straightforward.

Richard F. Gunst
Robert L. Mason
Tsushung A. Huu

iii

CONTENTS

Classroom Supplement to
REGRESSION ANALYSIS
AND ITS APPLICATION

CHAPTER 1

INTRODUCTION

Chapter 1 emphasizes the necessity for becoming as familiar with a project as time and resources permit. Before initiating a formal regression analysis the research goals should be examined to determine (i) whether a regression analysis is appropriate, (ii) any limitations which might be inherent in the (proposed) data base, and (iii) the intended use of the conclusions that are to be drawn. By discussing the issues raised below, the reader should better appreciate the importance of these preliminary examinations of research projects.

1.1 EXERCISES

1.1.1 Applied

1. Name at least three undesirable consequences that can occur in a regression analysis if data-base problems go undetected or are ignored.

2. List potential sources of bias from opinion polls that are solicited by telephone.

3. A theoretical model can be defined from scientific principles or arguments which are external to any data collection or analysis. Define two theoretical models for population growth.

4. An empirical model is one which is defined on the basis of observational evidence; for example, from previous data analyses. Is an empirical model or a theoretical one more appropriate for predicting population growth in the United States?

5. With reference to the Mortality and Pollution Study, suppose mortality rates are calculated for each of k age groups: M_1, M_2, M_3,..., M_k. Due to differences in population distributions it is sometimes desirable to "standardize" the age-specific mortality rates (the M_j) for a hypothetical common population with N_1, N_2, N_3,..., N_k persons in the individual age groups. Construct an "age-adjusted" mortality rate using the M_j and the N_j. (Note that age-adjusted mortality rates are not uniquely defined, this is just one example.) Under what general conditions would this mortality rate be superior (inferior) to an unadjusted rate (i.e., simply the total number of deaths per 100,000 population without any consideration of age)?

1.1.2 Theoretical

1. Which of the following regression models are linear?

   (a) $Y = \alpha + \beta_1 X^{\beta_2} + \epsilon$

(b) $Y = \alpha + \beta_1 e^x + \beta_2 \sin(X) + \beta_3 \ell n(X) + \epsilon$

(c) $Y = \alpha X^\beta \epsilon$

(d) $Y = \alpha + \beta_1 X + \beta_2 X^2 + \beta_3 X^{-1} + \epsilon$

Can any of the nonlinear models be transformed to linear ones? Which ones (if any)?

2. Give an example (other than the one in *RAA*) where correlation analysis is preferred to regression analysis.

3. Explain how, at times, good prediction can result even though a model is misspecified.

4. What is the proper role of regression analysis in causal studies?

1.2 PROJECTS

1.2.1 *Solid Waste Analysis*

Golueke and McGaukey (1970)* state that the two principal objectives of their research are "to develop a comprehensive, factual basis for the quantitative aspects of the regional wastes management system by giving due consideration to the spatial and functional nature of the region" and "to use this factual base as the primary source of data in analyzing and estimating future volumes of solid wastes that would be handled within the system." They define four "major variables" which explain the generation of solid wastes (i.e., garbage, rubbish, street refuse, etc.): population, income, employment, and land use. Data Set B.2 in *Regression Analysis and Its Application (RAA)* presents information only on the last of these categories of variables; viz., land use. (Note: We chose only to include the land use variables in order to keep the analysis manageable for study by the reader.)

Data on solid wastes were collected from nearly 100 waste disposal sites in the California Bay Area. Examination of the data collected convinced the researchers that only a fraction of the total amount of solid waste produced in the region was being disposed of at the disposal sites, approximately 35-40% of the total amount. The remainder (commercial, industrial, agricultural) was being disposed of elsewhere. The authors concluded that "The incompleteness inherent in data based solely on wastes received at disposal sites made it necessary to develop methods for making estimates of the total wastes generated by various sources before the analyses required by the present study could be made."

Ultimately the researchers categorized sources of solid wastes (e.g., major sources are residential, commercial, public agencies, industrial, and agricultural sectors of the region) and determined multiplication factors for each of the sources from other ongoing research studies. The amount of solid waste produced in a region was then estimated by determining the number of source units in the region (e.g., the number of single-family dwellings) and multiplying by the factors. For example, from a sample of 359 single-family dwelling units, the average quantity of rubbish produced was estimated at 2858 pounds per dwelling unit per year. This factor was then

---

*References which are identified only by author(s) and year of publication are completely listed in the Bibliography to *Regression Analysis and Its Application: A Data-Oriented Approach.*

2

multiplied by the number of single-family dwelling units in each region to estimate the total amount of rubbish generated by single-family dwellings in the respective regions. Similar techniques were employed on other solid waste sources to generate the total amount of solid waste produced in each of the 40 regions listed in Data Set B.2.

1. Although these exercises treat the solid waste variable as an actual measurement, it is an estimate from an economic model. What implications does this have on the intended goals of the study? (Note: the researchers recognized the problems associated with the generation of solid wastes and modified their analysis accordingly.) What practical problems necessitate this approach?

2. Our intentional exclusion of all predictor variables except land use variables could result in the regression model being misspecified. Under what general conditions would prediction of solid waste using only the land use variables be considered appropriate?

3. From the introduction to Data Set B.2 given above and in *RAA*, what limitations restrict inferences which can be drawn from an analysis of the data?

4. Assume an appropriate fit can be obtained for these data. Give an example for which inferences drawn on the fitted model would be considered extrapolations? generalizations? causal? Relate your answers to the results of Exercises 2 and 3.

1.2.2 Mortality and Pollution Study

An excellent discussion of the issues involved in a study of the potential impact of air pollution on mortality is contained in the following papers by L. B. Lave and E. P. Seskin:

"Air Pollution and Human Health," *Science, 169* (August 1970), 723-733.

"An Analysis of the Association Between U.S. Mortality and Air Pollution", *Journal of the American Statistical Association, 68* (June 1973), 284-290.

"Epidemiology, Causality, and Public Policy," *American Scientist, 67,* (1979), 178-186.

The data for this project are contained in Data Set B.1 and are taken from McDonald and Ayers (1978). Originally the data were analyzed in McDonald and Schwing (1973).

1. Section 1.2.1 of *RAA* briefly outlines some of the conceptual difficulties with a study of this nature. Read one or more of the above articles for a more in-depth appraisal of these difficulties. Summarize some of the most important issues raised in these articles.

2. Design an experiment using laboratory animals which would be satisfactory for verifying causal assumptions for the effects of air pollution on mortality. Can such an experiment be conducted to assess the effects of air pollution on human health? If so, carefully describe the experiment; if not, explain in detail why it cannot be conducted.

3. List some potentially important categories or classes of variables which have not been included in Data Set B.1. Define alternative (better?) choices for some of the predictor variables (e.g., POPN).

4. The mortality rate used in this study is a function of individual (age-adjusted) mortality rates which were calculated for several sex/race categories in each SMSA. Through this type of a computation the mortality rates are standardized for a common age distribution in each sex/race category. If the age distributions for these sex/race categories differ for several of the SMSAs, might this information be important to the study? Why (not)?

3

CHAPTER 2

INITIAL DATA EXPLORATION

Chapter 1 stressed the need for carefully examining the research problem prior to formally fitting regression models. Chapter 2 also stresses preliminary investigations of the regression problem but the emphasis is on exploration of the data base. One should bear in mind that these investigations are preliminary and must be supplemented with more advanced techniques which are detailed in subsequent chapters. Nevertheless, they are complementary to those techniques and provide basic insight which often is not available otherwise.

Two specific facets of initial exploration of a data base are detailed in this chapter: (i) inspection of the individual values in the data base and (ii) techniques for specifying an initial functional form for the response and predictor variables. The latter purpose is important when no theoretical model which defines the exact relationship between the response and the predictor variables is known.

2.1 EXERCISES

2.1.1 Applied

1. Calculate arithmetic averages of successive groups of three first-year average scores shown in Table 2.6. Compare these moving averages with those from the three-point median smoothing. Are any of the smoothed points substantially different for the two types of smoothing? State one advantage of each type of smoothing over the other.

2. Let $Y = 5 + 10X_1 + 2X_2 + 3X_1X_2$. Suppose X_2 is an indicator variable; i.e., X_2 takes on the values 0 and 1. Graph Y vs. X_1 for, say, X_1 between 0 and 10 and for each value of X_2. Repeat the exercise with the model $Y = 5 + 10X_1 + 2X_2$. What differences occur in the two graphs?

3. Construct scattergrams for NOX vs. HUMID, TEMP and BPRES for the emissions data in Data Set A.5. Also construct scattergrams for NOX vs. the square of each of these three predictor variables. Compare the two sets of graphs. Why are they so similar?

4. Carefully study the 44 observations on BPRES in Data Set A.5. Do any of the data values seem unusual? Which ones (if any)?

2.1.2 Theoretical

1. Superimpose on a graph plots of the functions $Y = X^2$, $Y = X^3$, and $Y = \exp(X)$ for X between 0 and 3. Observe that polynomial or exponential transformations using functions such as these might be appropriate for the data in Figure 2.2. Which transformations of

4

Y yield the same curves as those given above?

2. Using the functional relationships indicated in Figure 2.8 linearize the response for each of the graphs by plotting the inverse of the transformations.

3. The smoothing technique described in Section 2.2 of *RAA* is generally termed three-point median smoothing. The smoothing technique used in Exercise 1 (Applied) is termed a moving average. Would an outlier tend to affect moving average smoothing more than median smoothing? Why (not)?

4. Prove that use of p predictor variables and all their interactions in a model results in 2^p terms being included in the model.

5. Consider the two categorical variables

$$X = \begin{cases} 0 & \text{, if Category 1} \\ 1 & \text{, if Category 2} \end{cases}$$

and

$$W = \begin{cases} k_1 & \text{, if Category 1} \\ k_2 & \text{, if Category 2} \end{cases}$$

where k_1 and k_2 are any constants. Find values of α and β which produce the same response values in the model

$$Y = \alpha + \beta W + \epsilon$$

as does α^* and β^* in the model

$$Y = \alpha^* + \beta^* X + \epsilon.$$

2.2 PROJECTS

2.2.1 *Solid Waste Analysis*

1. Examine the individual values listed in Data Set B.2. Which of the observations appear to have unusual values? Can any of the observations be declared outliers? Why (not)?

2. Calculate summary statistics (e.g., max., min., mean, standard deviation) for each of the variables in the data set. Use these statistics to aid in your assessment of the individual data values. Does this information reinforce the impressions obtained by scanning the data?

3. Construct scatterplots of the response variable with each of the predictor variables. Which of the conclusions that were drawn in the previous two exercises do these plots reinforce or contradict?

4. Do the scatterplots indicate the need for preliminary transformations of any of the variables? If so, what transformations are suggested by the plots?

5. Based on the conclusions drawn from the previous analyses, identify observations (if any) which should be carefully scrutinized as possible outliers. Second, respecify the assumed model, if needed. (Once again, these are only preliminary judgements which might be altered later.)

2.2.2   Mortality and Pollution Study

As an aid to the interpretation of this regression analysis, the identification of the 60 Standard Metropolitan Statistical Areas (SMSAs) in Data Set B.1 is included in this Supplement as Appendix A. Reference to the SMSAs is important to the selection of appropriate answers to the following exercises.

1.   Using the identification of the SMSAs in Appendix A suggest some states or regions of the U.S. which might be underrepresented or overrepresented. How does this information impact possible conclusions which might ultimately be drawn following an analysis of the data set?

2.   Repeat Exercises 1, 2, 3, and 4 in Section 2.2.1 (above) for the data contained in Data Set B.1. Which of the observations appear to cluster in extreme regions of the scatterplots? Using the identification of the SMSAs in Appendix A, suggest a physical explanation for the clusters of extreme points (especially in the plots of MORT vs. HC and NOX).

3.   Extreme observations which have been identified fall into two categories: (i) possible individual outliers, (ii) clusters of points which necessitate special treatment due to a common identifiable physical characteristic. Categorize each of the observations which appear to have extreme values into one of these classes.

4.   Observations which cluster or group because of a physically definable characteristic can be incorporated in a regression model by adding one or more predictor variables to account for the characteristic. Suggest an appropriate redefinition of the model which will account for any clusters occurring in this data set. (Caution: if clusters only exist for certain variables, this fact also might need to be incorporated in the initial specification of the model.)

5.   Repeat Exercise 5 (above) for this data set.

6

CHAPTER 3

SINGLE-VARIABLE LEAST SQUARES

Introductory in nature, this chapter begins discussion on model specification and estimation of model parameters. Single-variable regression models are covered in a separate chapter from multiple-variable models for two reasons: (i) single-variable models are widely used and of special importance due to their relative simplicity, and (ii) certain concepts can be more readily demonstrated with single-variable models than with multiple-variable ones. Discussions of model specification, model assumptions, and estimation will continue throughout the remaining chapters of the text, especially in Chapters 5 and 6. Those later discussions build on the principles established in this chapter; indeed, the issues raised with single-variable models are equally valid with more complex models.

3.1 EXERCISES

3.1.1 Applied

1. Data Set A.8 contains body measurements for 33 individuals. Conceptually, if an individual's sitting height (SITHT) is zero her height must also be zero. Fit both an intercept and a no-intercept model for the prediction of height from sitting height. Which of these two fits appears better for this data set? Why?

2. Fit $(AGE-52)^2$ to the female suicide rates in Table 3.4. How do measures of the adequacy of the fit (e.g., R^2, $\hat{\sigma}^2$) compare with the linear fit in Exercises 10 and 11 of *RAA*?

3. As an indication of the amount of redundancy in LIC and GR for Data Set A.3, (i) make a scatter plot of these two variables, (ii) regress standardized values of LIC on standardized values of GR, and (iii) calculate measures of the adequacy of the fit. Explain how each of these techniques identifies redundancies in the variables.

4. Calculate the correlation coefficient between the grade 13 scores and the first-year college averages for high school 1 in Data Set A.2. Also calculate the means and standard deviations for each of these two variables. From only these quantities and the sample size, calculate the least squares coefficient estimates for the regression of the first-year college averages on the grade 13 averages.

3.1.2 Theoretical

1. Show that the least squares estimator given by eqn. (3.2.10) minimizes the sum of squared residuals for the no-intercept model.

7

2. Derive the relationships (e.g., Appendix 3.B) between the coefficient estimates for the original and standardized prediction equations when
   (i) both the predictor variable and response variable are standardized using eqn. (3.2.11).
   (ii) the predictor variable only is standardized using eqn. (3.2.14).
   (iii) both the predictor variable and response variable are standardized using eqn. (3.2.14).

3. Verify the changes that were suggested on p. 72 of RAA to $\hat{\alpha}$ and $\hat{\beta}$ when (i) X is replaced by aX + b, (ii) Y is replaced by cY + d, and (iii) both transformations are applied.

4. Using Pearson's product moment correlation coefficient show that the correlation coefficient between the residuals and the predictor variable is zero.

5. Show that when the estimated slope coefficient is zero the estimate for σ^2 given in eqn. (3.3.8) reduces to a multiple of the sample variance, S^2, of the response variable.

6. Algebraically establish the relationship between the beta weight and the correlation coefficient.

7. Show that the least squares estimator $\hat{\beta}$ can be written as a weighted average $\Sigma \omega_{ij} \hat{\beta}_{ij} / \Sigma \omega_{ij}$ of all 2-point slopes $\hat{\beta}_{ij} = (Y_i - Y_j)/(X_i - X_j)$ between pairs of points in a data set, where $\omega_{ij} = (X_i - X_j)^2$. Numerically verify this property using the $\hat{\beta}_{ij}$ shown in Table 3.2.

3.2 PROJECTS

3.2.1 *Solid Waste Analysis*

1. Rather than assess the impact of several land use variables on the amount of solid waste produced in a region, a simplification of the problem could be possible. Researchers often combine several similar variables into a single composite variable with the intent to reduce the complexity of the analysis. Consider defining a variable "TOTAL" to be the total land use for all sources listed in Table B.2; i.e.,

   TOTAL = INDUS + METAL + ... + HOME.

   How does the use of TOTAL instead of the individual land use variables alter the achievable goals of this study?

2. What are possible advantages associated with using TOTAL instead of the individual land use variables? What are possible disadvantages?

3. Construct a scatterplot of WASTE vs. TOTAL. What features of this plot reflect the conclusions drawn in response to Exercise 5 of Section 2.2.1? Are any important characteristics of the plots of WASTE with the individual land use variables not apparent in the plot of WASTE vs. TOTAL?

4. Calculate measures of the adequacy of the fit of WASTE to TOTAL. Is the fit acceptably accurate? Why (not)?

5. Regardless of the answer to the previous exercise, could an acceptable fit to this data prove the validity of the assumed model? Why (not)?

6. Calculate the beta weights for the regression of WASTE on TOTAL when (a) observa-

tion 2 is removed and only the remaining 39 data points are used, and (b) observation 8 is removed and the other 39 data points are retained. Use the scatterplot of WEIGHT vs. TOTAL to aid in explaining differences between these estimates and the estimate from the complete data set.

3.2.2 Mortality and Pollution Study

1. Suppose a researcher desires to study the effect of only one pollution variable on the mortality rate; e.g., HC, NOX, or SO2. Why would an analysis using only the individual predictor variable under study generally be inadequate?

2. Suppose, hypothetically, a regression analysis of MORT on HC produced a large R^2 and other measures of the adequacy of the fit were also good. What specific conclusions could be drawn relative to the influence of HC on MORT?

3. Continuing the previous exercise, calculate the correlation coefficient between the 40 values of HC and NOX in Data Set B.2. Would the hypothetical conclusions previously drawn on the association of HC and MORT be changed by knowing the value of the correlation between HC and NOX? Why (not)?

4. How does the magnitude of the correlation coefficient between HC and NOX affect any conclusion which one might draw on the use of this data set to support a causal assumption on the effect of HC on MORT?

CHAPTER 4

MULTIPLE-VARIABLE PRELIMINARIES

The intent of this chapter is to provide a foundation from which the reader can effectively study multiple linear regression analysis. The first section contains an overview of vector and matrix algebra, topics which can be briefly reviewed by readers who are already familiar with the material presented. The following three sections expand on problems which can arise through the specification of the predictor variables, notably those associated with predictor variable redundancies. Standardization of the predictor variables is advocated to remove numerical difficulties arising from different measurement scales for the individual variables. Examination of pairwise correlations and the latent roots and latent vectors of $W'W$ is shown to be a valuable procedure for detecting and identifying multicollinearities among the predictor variables.

4.1 EXERCISES

The numerical results given in *RAA* for Data Set A.8 utilize BRACH and TIBIO measurements which have five decimal places rather than the two reported on p. 367 for Data Set A.8 in order to achieve better accuracy and obtain the same numerical results as appear in the text, compute BRACH and TIBIO as on p. 349 of *RAA* (and express them as a percentage with five decimal places).

4.1.1 Applied

1. Calculate $X^{*'}X^*$, $W'W$, $(W'W)^{-1}$ and the latent roots and latent vectors of $W'W$ for a prediction equation which uses only the two predictor variables ULEG and LLEG from Data Set A.8.

2. Repeat Exercise 1 for the three predictor variables: ULEG, LLEG, and TIBIO. What multicollinearities, if any, are observable from these measures?

3. Consider the first five anthropometric measurements given in Data Set A.6 (i.e., HEIGHT, WEIGHT, SHLDR, PELVIC, and CHEST) and the response variable REACT. Identify any strong multicollinearities that are present by examining the pairwise correlations among the predictor variables and the latent roots and latent vectors of $W'W$.

4. Consider Data Set A.4 where EDATT is the response variable. Are any multicollinearities present in this data base? Which ones (if any)?

4.1.2 Theoretical

1. Write $X^{*'}X^*$ and $(X^{*'}X^*)^{-1}$ for single-variable regression models in terms of quantities

such as ΣX_i and ΣX_i^2.

2. Using the results of Exercise 1, express $(X^{*\prime}X^*)^{-1}X^{*\prime}\underline{Y}$ in terms of ΣX_i, ΣX_i^2, ΣY_i, $\Sigma X_i Y_i$. Show that the solution is equivalent to eqn. (3.2.6).

3. Suppose in a single-variable regression model the mean, $\bar{X}$, of the predictor variable is zero. How does this effect the quantities derived in Exercises 1 and 2? What are the latent roots and latent vectors of $(X^{*\prime}X^*)^{-1}$ in this case?

4. What characteristic would the predictor variable have to exhibit in order for a strong multicollinearity to exist in a single-variable model?

4.2 PROJECTS

4.2.1 Solid Waste Analysis

1. Using an appropriate computer program, compute the pairwise correlations between all pairs of predictor variables in the original data set. Which (if any) correlations indicate strong multicollinearities? Do the correlations reveal any peculiar associations among the last five predictor variables (RETAIL, REST, FINAN, MISC, HOME)?

2. Compute the latent roots and latent vectors of the matrix of correlations of the predictor variables; i.e., of $W'W$. How is the large pairwise correlation between RETAIL and MISC identifiable in the latent vector corresponding to the smallest latent root of $W'W$?

3. Underline all elements of $\underline{V}_1$ and $\underline{V}_2$, the latent vectors corresponding to the two smallest latent roots of $W'W$, which are at least 0.40 in magnitude. Which predictor variables do these large elements identify? How does this finding relate to the last answer in Exercise 1 above?

4. Vectors $\underline{V}_1$, $\underline{V}_4$, and $\underline{V}_5$ clearly show two large elements which are almost equal in magnitude and are opposite in sign. The latent roots corresponding to these latent vectors differ considerably. Interpret these results in terms of the nature and the strength of the respective multicollinearities.

5. If one desires to assess which of the land use variables are most closely associated with large volumes of solid waste production, how do the above results impact any conclusions which might ultimately be drawn?

4.2.2 Mortality and Pollution Study

1. This data set contains only one strong multicollinearity. Identify it by examining both the pairwise correlations among the predictor variables and the latent roots and latent vectors of $W'W$.

2. Construct an indicator variable CALIF which equals 1 if the SMSA is from California and zero otherwise. Next construct two interaction terms HC*CALIF and NOX*CALIF. If these latter two interactions are added to the original 15 predictor variables in order to adjust for the extreme points observed in the plots of MORT vs. HC and NOX, what multicollinearities occur in the enlarged data set?

3. As one possible alternative to the above interactions, consider adding only the indicator variable for the California SMSAs. Would this variable account for the tendencies

observed in the scatterplots? Why (not)?

4. Other alternatives to the addition of the above two interaction terms are: (i) add CALIF, HC*CALIF, NOX*CALIF, and S02*CALIF, or (ii) add an indicator variable for each California SMSA. Why might these alternatives be inferior to the inclusion of the two interaction terms? (See also Exercise 3, Section 5.2.2, and Exercise 5, Section 7.2.2.)

CHAPTER 5

MULTIPLE-VARIABLE LEAST SQUARES

With the understanding that further adjustments on the assumed model might be required, this chapter discusses the fitting of multiple-variable regression models. Two approaches are developed for estimating regression coefficients. The matrix-theory approach is presented because distributional properties and variable selection techniques are readily derived when the estimators are expressed in matrix form. The second approach to the estimation of regression coefficients is "fitting-by-stages". Conceptually this approach points out that estimated coefficients are actually produced by adjusting a predictor variable for the common information it shares with other predictor variables. Estimated coefficients therefore actually measure only the unique contribution of individual predictor variables to the prediction of the response variable.

In addition to the mechanics of estimating regression coefficients, this chapter stresses the interpretation of the estimates and it introduces a few measures of the accuracy with which the fitted model predicts the observed responses. The Analysis of Variance Table is algebraically derived and is used frequently in this and subsequent chapters to assess the adequacy of the fit.

5.1 EXERCISES

5.1.1 Applied

1. Construct an ANOVA table to accompany the prediction of EDATT using the raw predictor variables in Data Set A.4. What specific information cannot be determined from only summary data such as Data Set A.4?

2. Can the estimated regression coefficients for Data Set A.4 be individually interpreted?

3. Calculate the (unit-length) standarized coefficient estimates for Data Set A.4. Use the magnitudes of the estimates to draw preliminary conclusions regarding the relative influence of the predictor variables. Which predictor variable appears to most influence the prediction of EDATT? From the definitions of the predictor variabies does this result appear reasonable? Why (not)?

4. Does the foregoing analysis alleviate any of the fears which were raised in the third paragraph on p. 130 of *RAA*? Why (not)?

5. Construct the ANOVA table for the police applicant data, Data Set A.8. Does the summary information in the table indicate an acceptable fit to the heights of these police department applicants?

6. Calculate the standardized estimates for the police applicant data. Note in particular the large magnitude for LLEG. Can one conclude that this predictor variable has the greatest influence on the response variable?

5.1.2 Theoretical

1. Show algebraically that the predicted responses using the unit length standardization are identical to those using the raw predictor variables.

2. Show that when the first r columns of X^*, denoted X_1^*, are orthogonal to the remaining $(p+1) - r$ columns of X^*, denoted X_2^*, the corresponding sets of parameters in $\underline{B}$, say $\underline{B}_1$ and $\underline{B}_2$, are separately estimated as indicated on p. 141 of *RAA*.

3. Verify that, when X_1^* is not orthogonal to X_2^*, the least squares estimator of $\underline{B}_2$ is given by

$$\hat{\underline{B}}_2 = [X_2^{*\prime} X_2^* - X_2^{*\prime} X_1^* (X_1^{*\prime} X_1^*)^{-1} X_1^{*\prime} X_2^*]^{-1}$$
$$\cdot [X_2^{*\prime} \underline{Y} - X_2^{*\prime} X_1^* (X_1^{*\prime} X_1^*)^{-1} X_1^{*\prime} \underline{Y}]$$

[Hint: Follow the steps described in Appendix 5.B of *RAA* particularly those leading to eqns. (5.B.1) and (5.B.2)].

4. Show algebraically that SSR is invariant (numerically unchanged) if one uses the normal deviate transformation of the predictor variables.

5. Show that SSR is invariant to any transformation $X^{**} = X^*T$, where T is a $(p+1) \times (p+1)$ nonsingular matrix.

6. Verify algebraically eqn. (5.3.5) as corrected in Appendix C for the beta weight standardization.

5.2 PROJECTS

5.2.1 Solid Waste Analysis

1. Construct an ANOVA table for the regression of WASTE on SQRTIN (square root of INDUS), SQRTME (square root of METAL), WHOLE, REST, FINAN, RETAIL, MISC and HOME. Make an initial assessment of the adequacy of the fit. Discuss why the fit is (is not) adequate.

2. Construct an ANOVA table for a fit of the original data set using INDUS and METAL. Do the ANOVA tables for this fit and the previous one clearly indicate that one of the two sets of variables is preferable to the other? Why (not)?

3. Compare the two previous fits with the single-variable fit of WASTE to TOTAL. Is there a clear indication, on the basis of the ANOVA tables alone, that one or two of the fits is (are) superior to the other(s). Why (not)?

4. Calculate the estimated regression coefficients for an eight-variable prediction equation for WASTE using SQRTIN, SQRTME, WHOLE, RETAIL, REST, FINAN, MISC and HOME. Compare the relative magnitudes of the raw coefficient estimates with those of the standardized (unit length or beta weights) estimates, e.g., compare the relative magnitudes of the coefficient estimates for SQRTME and WHOLE. Why do some of the relative magnitudes change for the two sets of estimates? Which set is preferable for comparison purposes? Why?

14

5. Identify strong multicollinearities in the transformed data set. Observe, for example, that, SQRTIN is not strongly multicollinear with any of the other predictor variables, and that RETAIL is highly correlated with other predictor variables, especially with MISC. How does this finding affect the interpretation of the magnitudes of the estimated coefficients (i.e., the interpretation of the influence of each predictor variable on the response as measured by the magnitude of the coefficient estimates).

5.2.2 *Mortality and Pollution Study*

1. Examine the ANOVA tables for the fit of the original 15 predictor variables to MORT and the fit for the 17 variable prediction equation in which CALIF*HC and CALIF*NOX are added to the original variables. On the basis of the information in the ANOVA table is one of the fits clearly superior to the other? Why (not)?

2. Make two alternative fits for this data set. To the original 15 predictor variables add (a) four indicator variables, one for each of the California SMSAs, and (b) CALIF, CALIF*HC, CALIF*NOX, and CALIF*SO2. Which, if any, of the three fits constructed thus far appears superior?

3. The ANOVA tables for the two fits in the previous exercise are identical. Why does this occur?

4. Suppose that in Exercise 2(a) one or more of the indicator variables for the California SMSAs has a large coefficient estimate. What does this imply about the fit to the data?

CHAPTER 6

INFERENCE

Statistical tests of hypotheses and interval estimation are the two major topics introduced in this chapter. With the ability to perform tests of hypothesis and calculate interval estimates much of the subjectivity of previous analyses can be removed. In addition, more intense focus can be directed toward specific questions of interest to the researcher such as whether a particular predictor variable is an important contributor to the prediction of the response. Prior to performing tests of hypothesis and determining interval estimates it is necessary that model assumptions, including the specification of the variables, be critically examined. This chapter defines a series of model assumptions and addresses their impact on the various procedures which are presented. The next chapter stresses verification of the model assumption.

6.1 EXERCISES

Data Set A.5, Nitrous Oxides Emissions Modelling, is reproduced in this Supplement in Appendix A. In this appendix HUMID and TEMP are reported to two decimal places rather than one as in *RAA*. In addition, some minor corrections are made to a few of the observations. Use the data set in Appendix A for all exercises in the text and in the supplement that involve Data Set A.5.

6.1.1 Applied

1. Test the hypothesis $H_0:\beta_1 \geq 0.1$ versus $H_a:\beta_1 < 0.1$ for the three variable prediction equation of NOX on $X_1 = $ HUMID, $X_2 = $ TEMP, and $X_3 = $ BPRES.

2. Construct 95% confidence limits for $E[Y]$ for the average annual salary data given in Table 2.1 of *RAA*. Plot the confidence intervals on a graph similar to Figure 6.5.

3. Fit the four variable prediction equation of ACR on CNHF13 (X_1), RES510 (X_2), LTOTARE (X_3), and QUAL2 (X_4) from Data Set A.7. Test the appropriate hypothesis for determining if CNHF13 is an important predictor variable in this model.

4. Construct an ANOVA table for the fit in Exercise 3. Test the hypothesis $H_0:\alpha^* = 0$ vs. $H_a:\alpha^* \neq 0$ and $H_0:\underline{\beta} = \underline{0}$ vs. $H_a:\underline{\beta} \neq \underline{0}$.

5. Fit NOX to HUMID and TEMP in Data Set A.5. Construct individual 95% confidence intervals for the unknown coefficients of HUMID and TEMP as well as a 95% joint confidence region. Plot these three results on a HUMID vs. TEMP graph. What difference is noticeable between the joint confidence region and the area formed by the two individual confidence intervals?

16

6.1.2 *Theoretical*

1. For a single-variable regression model derive the variances of the least squares estimators of α and β.

2. State the modifications needed in Tables 6.4 and 6.5 to allow for performance of one-sided tests of hypothesis.

3. Show that the least squares estimators for the normal deviate prediction equations are unbiased for their respective parameters. Show also that transforming these standardized estimators back to estimators of the original model parameters produces unbiased estimators of the elements of $\underline{B}$.

4. Consider the following interaction-term model

$$Y = \alpha + \beta_1 X_1 + \beta_2 X_2 + \beta_3 X_1 X_2 + \epsilon$$

   where X_1 is numerical and X_2 is categorical with values 1 and 0. Since models containing interaction terms yield regression lines having different intercepts and slopes for each value of the categorical variable, two regression lines have been combined into a single model. State (in terms of β_1, β_2, and β_3) a hypothesis which is appropriate for testing (i) for parallelism of the two lines, (ii) for a common intercept, and (iii) for coincidence of the two lines.

5. Let $Y_i = \alpha + \beta_1 X_{i1} + \beta_2 X_{i2} + \epsilon_i$ and let r_{Y1}, r_{Y2}, and r_{12} denote the correlation coefficients between Y and X_1, Y and X_2, and X_1 and X_2, respectively. Show that if $r_{12} = 0$, (a) $\hat{\beta}_j^0 = r_{Yj}$, and (b) $R^2 = r_{Y1}^2 + r_{Y2}^2$. Generalize these results to p predictor variables.

6.2 PROJECTS

6.2.1 *Solid Waste Analysis*

1. From the descriptions of the land use variables given earlier in this Supplement and in the introduction to Data Set B.2, discuss the validity of Assumption 2, Table 6.2. In light of your answer to this exercise, would Assumptions 2 and 4 or 3′ and 4′ be more reasonable for this data set? Why?

2. Perform a test of hypothesis for the significance of the regression coefficients using SSM and SSR from the ANOVA table for the fit of the transformed predictor variables (see Exercise 1, Section 5.2.1). Conduct similar tests for any other fits which appear to be reasonable alternatives from previous analyses. If differing conclusions are drawn for any of the tests, explain why they occur.

3. The raw coefficient estimate for HOME using the transformed predictor variables is over an order of magnitude smaller than that of MISC. The t (or F) statistic for HOME, however, is highly significant but that of MISC is not. Does the same juxtaposition occur between the (unit length) standardized coefficient estimates and the t statistics? How do these results relate to the comparison of coefficient estimates using raw vs. standardized estimates?

17

4. Construct a 95% confidence interval for the true coefficients of SQRTIN and MISC. Compare the lengths of the two confidence intervals: (a) does either coefficient appear to be estimated with sufficient accuracy? (b) what do the confidence intervals enable one to conclude about the *relative* accuracy of the estimates? What information do the intervals provide concerning the significance of the coefficient estimates?

5. Construct a 95% prediction interval for WASTE using the standardized (transformed) predictor variables and $u_j = \bar{X}_j$, $j = 1, 2, ..., 8$. Does this prediction interval indicate that the response can be predicted with a high degree of accuracy when each predictor variable equals its mean? Why (not)? Contrast this interval with a 95% confidence interval on $E[Y]$. Why is there such a large difference in the intervals?

6.2.2 *Mortality and Pollution Study*

1. List at least 3 reasons why the SMSAs in Data Set B.1 are not necessarily representative of all possible regions in the U. S. (e.g., locate the four California SMSAs on a map, notice that some states are unrepresented). How do these results affect Assumption 2, and consequently, Assumptions 3 and 4?

2. Examine the estimated (standardized) regression coefficients for HC and NOX, as well as the t (or F) statistics for testing the significance of these two coefficient estimates. Observe, in particular, the signs and the relative magnitudes. Are these estimates reasonable in light of the large positive correlation coefficient between HC and NOX?

3. Can the correlation between HC and NOX be considered inherent to the population of all SMSAs in the United States? Why (not)? Examine scatterplots and correlations between HC and NOX both with and without the California SMSAs. Does the lower correlation when the California SMSAs are removed reinforce or alter the above conclusion about whether the correlation is inherent to the population of SMSAs in the U.S.? Why?

4. In Exercise 2 an implicit a-priori assumption about the true regression coefficient values for HC and NOX was made. What is the assumption? Would the same assumption be obvious if the correlation between HC and NOX was unknown? If the answer to this question is yes, on what basis (if any) can the coefficient estimates be questioned (refer to the answer to Exercise 2)? If the answer is no, defend the reasonableness of the coefficient estimates in light of the large positive correlation between the two predictor variables.

5. Repeat Exercises 2 to 4 on CALIF*HC and CALIF*NOX in the expanded 17 predictor variable data set.

6. Compare the signs and relative magnitudes of the coefficient estimates for HC and NOX with the corresponding elements in $\underline{V}_1$ for the original 15 variable data set. What does this suggest about the relationship between these estimates and the multicollinearity between HC and NOX? Perform the same analysis on the expanded data set. Are the signs and relative magnitudes of the coefficient estimates consistent with those of the cor-

responding elements in the latent vectors of $W'W$ which identify the multicollinearities between these variables?

7. Make a preliminary assessment of the effect of the pollution variables for the original 15 variable data set. Does it appear that the pollution variables are providing information on the prediction of MORT over and above that provided by the other variables in the prediction equation? Why (not)?

CHAPTER 7

RESIDUAL ANALYSIS

Combined with the results of previous chapters, the techniques discussed in this chapter enable one to make a final assessment of model specification and outliers. The sensitivity of residual and partial residual plots sometimes suggests predictor variable transformations which are not noticeable in two-variable plots. Likewise, error assumptions can be evaluated with residual plots and normal probability plots. Once model specification questions have been answered, the residual statistics proposed in the last section of this chapter can be used to make a final determination about outliers.

7.1 EXERCISES

7.1.1 Applied

1. Examine the nine-variable prediction equation for Data Set A.8. Determine if the residuals from the fitted model reveal any apparent outliers or model inadequacies.

2. Study the partial residual plots resulting from the prediction of first year college grade averages using the predictions from eqn. (2.2.1). Is the need for a quadratic term in the grade 13 averages as evident as in Figure 7.8?

3. Compute leverage values, studentized residuals, deleted residuals, and studentized deleted residuals for the grade data in Exercise 2. Do these values indicate the presence of any apparent outliers? Explain.

4. Plot raw residuals versus deleted residuals for the grade data in Exercise 2. Are the results of this analysis consistent with those of Exercise 3?

5. The observations in Data Set A.5 are listed in the order in which the respective experiments were run. In experimental situations in which observations are taken sequentially, serial correlation sometimes arises. Perform the runs test and Durbin-Watson test on the residuals of the regression of NOX on HUMID, TEMP, and BPRES. What are your conclusions?

7.1.2 Theoretical

1. Verify that the elements of the H matrix satisfy the two conditions given in eqn. (7.1.6). (Hint: Show that $H = H^2$ and express h_{ii} in terms of the elements of H^2.)

2. Calculate the elements of the hat matrix for a single-variable regression model.

3. One type of residual plot that should be avoided is that of r_i versus the observed

responses Y_i since these variables are generally linearly correlated. Algebraically show that the correlation between residuals and observed responses is not, in general, zero.

4. Cook's distance measure, D_i, compares the estimated coefficients, $\hat{\underline{B}}$, of the full model with the estimated coefficients when the ith observation is removed, $\hat{\underline{B}}_{(-i)}$. Express this difference in terms of h_{ii} and the raw residuals, r_i.

5. Show that the distance measure D_i given in eqn. (7.4.2) can be expressed in the forms given in eqns. (7.4.3) and (7.4.4).

7.2   PROJECTS

7.2.1   Solid Waste Analysis

1. Compare the partial residual plots of INDUS and METAL from the fit to the original 8 predictor variable model (n = 40) with those of SQRTIN and SQRTME from the fit to the transformed 8 variable model. Which of the two sets of plots is preferable? Why?

2. Examine all the plots of the residuals vs. individual predictor variables and all partial residual plots for the fit to the eight-variable transformed model. Do these plots suggest the need for further predictor variable transformations? If so, reformulate the assumed model until no further transformations are suggested.

3. Examine a normal probability plot for the fit finally selected in Exercise 2. Apart from possible departures of a few points is the plot reasonably straight? If not, transform the assumed model until all residual plots and the normal probability plot are acceptable.

4. Using the model assumed on the basis of the above considerations, calculate leverage values, studentized residuals, studentized deleted residuals, and Cook's distance measure for the 40 observations in the data set. Combined with the outlier assessments made in earlier chapters, determine observations which should be deleted from the data base.

5. Compare the standardized coefficient estimates (or beta weights) and the individual t (or F) statistics for the n = 40 prediction equation resulting from Exercise 3 with those from the reduced data set after the outliers in Exercise 4 are deleted. Does the magnitude or significance of any of the individual predictor variables change substantially when the outliers are deleted? If so, refer to appropriate plots to ascertain the reasons for the substantial changes which are observed.

6. After deleting the outliers which were identified in Exercise 4, reevaluate residual plots, partial residual plots, normal probability plots, etc. to determine whether any further adjustments are needed on the prediction equation. What, if any, further alterations are needed?

7.2.2   Mortality and Pollution Study

1. Carefully study the partial residual plots for the original 15 predictor variable data set and for the 17 variable data set containing the two interaction terms CALIF*HC and CALIF*NOX. How does the inclusion of the interaction terms affect the plots (e.g., HC, NOX)?

2. The partial residual plots for the 15 and the 17 predictor variable models indicate the

possible need for transformations of the pollution variables. Study carefully the curvature in these plots. Then replace HC, NOX, and SO2 with their natural logarithms. Also form the interaction of CALIF with ℓn(HC) and ℓn(NOX). For this new transformed prediction equation, compare the partial residual plots with those of the previous fits. Which do you prefer? Why?

3. Make any final transformations which are suggested by the residual plots, normal probability plots, etc. Does the ANOVA table indicate that the fit from the resulting prediction equation to these 60 SMSAs is adequate? Why (not)?

4. Once the final form of the prediction equation is determined, calculate the residual statistics for this fit. Using these statistics, plots, and other criteria you deem relevant, which SMSAs would you delete as outliers? Why?

5. To the original 15 predictor variables add an indicator variable which is 1 for Los Angeles and 0 for all the other SMSAs. Compare the t statistic for testing the significance of this indicator variable with the studentized deleted residual for Los Angeles from the 15 variable fit. Interpret the results.

CHAPTER 8

VARIABLE SELECTION TECHNIQUES

Once an acceptable fit to a data set has been obtained, it is often desirable to determine whether some predictor variables can be eliminated from the prediction equation. Alternatively, a very large number of predictor variables could be available for possible use in a regression model, but it might be unknown which ones are most beneficial. Even if a hypothesized theoretical model is available, it is often important to test whether the data support the assumed model. All these concerns relate to the general topic of variable selection.

This chapter, coupled with the tests derived in Chapter 6, provides a range of procedures for selecting predictor variables. In situations wherein a theoretical model is unknown and a large number of predictor variables are candidates for inclusion in the prediction equation, techniques such as forward selection and backward elimination are widely used to reduce the number of analyses needed to obtain an acceptable fit to the data. Several selection criteria such as R^2, F, and C_k are discussed as measures of the influence of sets of predictor variables.

8.1 EXERCISES

8.1.1 Applied

1. Perform a forward selection on the first seven predictor variables in Data Set A.8. Construct a summary table which lists at each stage of the analysis the variable added, the F statistic, the cumulative R^2 values and the C_k values. Which subset(s) provide acceptable fits to the data on the basis of the R^2, F, and C_k statistics?

2. Perform a backward elimination of the first seven predictor variables in Data Set A.8. Summarize each step as in Exercise 1. Which subset(s) is (are) most appropriate? Compare these results with those of the forward selection method.

3. On the basis of the results of the previous two exercises, does it appear that forward selection and backward elimination are sufficient to determine a "best" subset for this data set? Why (not)?

4. Perform a forward selection and backward elimination of Data Set A.8 using all the predictor variables. Compare the results. Is a "best" subset clearly identified by these analyses? Why (not)?

5. How would the results of Exercises 3 and 4 differ if a smaller significance level, say 0.05, had been used?

8.1.2 Theoretical

1. Why is the C_k statistic given in eqn. (8.1.7) not an unbiased estimator of Γ_k given in eqn. (8.1.8)?

2. Express C_k in eqn. (8.1.7) as a function of R_i^2 in eqn. (8.1.6) and R^2 for the full model in eqn. (8.1.2).

3. An alternative to R_i^2 as a measure of the adequacy of the fit of a reduced prediction equation is the adjusted coefficient of determination,

$$\bar{R}_i^2 = 1 - \left(\frac{n-1}{n-p-1} \right) \left(\frac{SSE_1}{TSS(adj)} \right)$$

$$= 1 - \left(\frac{n-1}{n-p-1} \right) \left(1 - R_i^2 \right)$$

Show by numerical illustration the advantages of using $\bar{R}_i^2$ over R_i^2.

4. In conjunction with a forward selection analysis it is often stated that the first variable that enters is the most important, the second that enters is the next most important, etc. Is this statement correct? Explain your answer.

5. Name three advantages and three disadvantages of using a forward selection method in a regression analysis.

6. Repeat Exercise 5 for backward elimination.

8.2 PROJECTS

8.2.1 Solid Waste Analysis

1. Using the transformed data with observations 2, 31, and 40 deleted, perform a forward selection of the predictor variables. Make a table which lists, at each stage of analysis, the variable added, the F statistic for judging whether to include the variable, and the cumulative R^2 and C_k values.

2. Compare the order of inclusion of variables with their pairwise correlations with the response variable. Why do the predictor variables not enter the prediction equation in the same order as the magnitudes of their correlations with WASTE? What insight does this result provide with respect to preliminary selection of predictor variables solely on the basis of their correlation with a response variable?

3. Select a subset of predictor variables using forward selection (a) on the basis of the R^2 values, (b) by comparing the F statistics with a suitable value from F tables, and (c) by examining a C_k plot. Which of the subsets is preferable? Why?

4. Repeat the above three exercises using backward elimination.

5. Determine one or two subsets which you could recommend as adequate from the foregoing analyses. Why was (were) this (these) subset(s) selected?

6. Conduct a t-directed search using the second selection procedure outlined in Section 8.2.4 of the text. Compare these results with those of forward selection and backward elimination. (Note: the stepwise procedure of Section 8.3.3 is, for this data set, identical with forward selection.) Does the conclusion to the previous exercise change on the basis of this analysis? Why (not)?

8.2.2   *Mortality and Pollution Study*

1. Focus on the first seven steps in a forward selection of the full ($n = 60$) transformed data set ($p = 17$). Identify the seven predictor variables which enter by stage 7. Do these results support a conclusion that the pollution variables are strongly associated with high mortality rates? Why (not)? Can one conclude, on the basis of this analysis, that the pollution variables are among the most important determinants of high mortality rates? Why (not)?

2. Complete the forward selection procedure and construct a summary table similar to the one requested in Exercise 1 in Section 8.2.1. Which subset(s) provide acceptable fits to the data on the basis of R^2, F, and C_k statistics?

3. Perform a backward elimination analysis of this data set. Carefully note how the step-by-step elimination of variables differs with the step-by-step inclusion with forward selection. Which of the backward elimination subsets appears most appropriate for this data set? Why?

4. Select variables for a final prediction equation using either a "best subset" algorithm or one of the t-directed search routines described in Section 8.2.4 of the text.

5. Of all the analyses performed thus far, which subset(s) appears to be most appropriate for this data set? Why?

CHAPTER 9

MULTICOLLINEARITY EFFECTS

Before final acceptance of the fit of the predictor variables to the response, an examination of multicollinearities (if any) in the data set should be conducted. If two predictor variables are highly correlated, for example, a variable selection procedure might indicate that one of the variables should be deleted. Knowing that the two variables are multicollinear offers one the possible option of deleting the other variable and retaining the one which is most beneficial for theoretical or economic reasons. Thus, prior to acceptance of the fit (including the estimates, results of variables selection, etc.) multicollinearities should be identified, their possible effects noted, and a remedial strategy determined.

9.1 EXERCISES

9.1.1 *Applied*

1. Reexamine the multicollinearities among the predictor variables ULEG, LLEG, and TIBIO in Data Set A.8 (see Supplement Exercise 2, Section 4.1.1). Calculate the least squares estimates and t statistics for the regression of HEIGHT on these three variables. Are the effects of the multicollinearities evident in the estimates and t statistics? Explain.

2. Contrast the results of the previous exercise with those from a two-variable prediction equation containing only ULEG and LLEG.

3. Reexamine the multicollinearities among the first five predictor variables in Data Set A.6 (see Supplement Exercise 3, Section 4.1.1). Are the effects of the multicollinearities evident in the estimates and t statistics for these predictor variables? Explain.

4. Regress ACR from Data Set A.7 on CNHF13, RES510, QUAL2 and LTOTARE. Compare the beta weights with those from the regression of ACR on these four predictor variables and LTOTRMS. Interpret any differences in the respective coefficient estimates.

9.1.2 *Theoretical*

1. Show that the jth diagonal element of $Q = (W'W)^{-1}$ is given by eqn. (9.1.1).

2. Many regression computer programs only enter a predictor variable, say X_j, into the regression equation provided it passes a "tolerance" limit; i.e., provided R_j^2 does not exceed 1.0 - TOLERANCE, where TOLERANCE is a small value chosen by the user. What is the relationship between this tolerance value and a variance inflation factor?

3. Consider a regression model containing only two predictor variables, $\underline{W}_1$ and $\underline{W}_2$ and let $r = \underline{W}_1'\underline{W}_2$ be the simple correlation between $\underline{W}_1$ and $\underline{W}_2$. Show that as $r^2 \to 1$, the $\mathrm{Var}(\hat{\beta}_1^*) \to \infty$ and $\mathrm{Cov}(\hat{\beta}_1^*, \hat{\beta}_2^*) \to \pm\infty$.

4. Express the variance of $\hat{Y}$ at a given standardized point, say $\underline{u}$, in terms of the latent roots and latent vectors of $W'W$. Show that if a single multicollinearity exists in the sample then $\mathrm{Var}[\hat{Y}]$ is not inflated due to the multicollinearity if $\underline{u}'\underline{V}_1 = 0$, but it can be very large if $\underline{u}'\underline{V}_1 \neq 0$.

5. One suggested solution to the multicollinearity problem is to collect more data, thereby removing the collinearity from the data base. In what direction of the predictor variable space should one collect such data in order to maximize the infuence of the additional points while minimizing the impact of the multicollinearity?

9.2 PROJECTS

9.2.1 Solid Waste Analysis

1. Calculate the variance inflation factors for the transformed solid waste data set with observations 2, 31, and 40 deleted. This can be accomplished (i) by writing a computer program, (ii) by using a computer program which will invert the correlation matrix of predictor variables, or (iii) from the standard errors of the raw coefficient estimates (see text Exercise 9.2) or from standardized coefficient estimates and their corresponding t statistics using eqn. (9.2.1). From the variance inflation factors identify predictor variables which are involved in multicollinearities.

2. From the correlation coefficients between pairs of predictor variables and the latent roots and latent vectors of $W'W$, identify the nature of the relationships among the multicollinear predictor variables.

3. Are there any obvious effects of the multicollinearities on the numerical values of the least squares estimates or t statistics in this data set? Explain.

4. Examine the step-by-step selection of variables in forward selection and backward elimination to determine whether one of a pair (or set) of multicollinear predictor variables was included in the prediction equation but not the other(s). Does this analysis suggest alternative subsets? Which ones (if any)?

5. Do the multicollinearities in this data set suggest further alternatives to the subsets of predictor variables identified in Exercises 5 and 6 in Section 8.2.1? Which ones? Evaluate these alternative subsets to determine whether R^2 or C_k change substantially when the substitutions are made. Which subsets now appear to be the best fits to this data set?

6. If the final subset(s) of variables selected in the previous exercise do not use all of the eight original predictor variables, reconsider whether points 2, 31, and 40 are outliers. Do scatterplots, residual statistics, etc. still dictate the removal of these observations? Why (not)?

9.2.2 Mortality and Pollution Study

1. Calculate variance inflation factors and latent roots and latent vectors for the seventeen-

variable transformed data set. Identify the number and nature of all the multi-collinearities.

2. Prior to the possible elimination of any variables from this data set, observe the effects of the multicollinearities on the coefficient estimates and individual t statistics (as well as the standard errors). Describe these effects.

3. Can one of the two interaction terms be eliminated from the data set because of their high correlation? Why (not)?

4. Reconsider the final subsets obtained using variable selection techniques in the last chapter. Do the multicollinearities suggest possible substitution of predictor variables? Which ones (if any)? Reanalyze the subsets for which substitutions are possible. What final subsets appear to be most appropriate for this data set?

CHAPTER 10

BIASED REGRESSION ESTIMATORS

Biased estimators provide an alternative to least squares estimators and test procedures. They offer the potential for a great reduction in the variance of estimators over the variances of the least squares estimators. Judiciously used, the bias incurred with these estimators can be very small relative to the great reduction in variance. Although this bias is always unknown, if multicollinearities are very strong the use of biased estimators is appropriate when (i) the reduction in variance is substantial and (ii) the effect of the multicollinearities on the numerical values of the least squares estimates is evident from their signs and magnitudes.

10.1 EXERCISES

10.1.1 Applied

1. Calculate the least squares estimates for the "standardized" emissions data (see Text Exercise 9.8). Use forward selection and backward elimination to obtain suitable subsets of the predictor variables.

2. Perform principal component regression analysis of the standardized emission data. Do the predictor variables with large t statistics correspond to the same subset of variables chosen by either of the variable selection procedures in Exercise 1?

3. Repeat Exercise 2 for the latent root regression estimator.

4. All three of the estimators in the previous exercises yield different coefficient estimates and different subsets of significant predictor variables. Select a few predictor variables which do not appear to have significant predictive value in any of the analyses. Eliminate these predictor variables and redo all three analyses. Are the results more consistent? Explain your answer.

5. Select ridge parameter values for the reduced data set in Exercise 4 (a) from a ridge trace and (b) such that the largest variance inflation factor is less than 10 (use multiples of .01 for trial parameter values).

6. Which of the analyses in Exercises 4 and 5 are most appropriate for this data set? Why?

10.1.2 Theoretical

1. Derive the principal component estimator by minimizing the residual sum of squares subject to the restrictions $V'_{CD}\hat{\beta}^* = 0$.

2. Using eqns. (10.1.7) and (10.1.8), show that

$$SSR = SSR_{PC} + SSR_{CD} \;,$$

where

$$SSR_{PC} = \hat{\beta}_{PC}^{*\,\prime} W' W \hat{\beta}_{PC}^* \text{ and } SSR_{CD} = \hat{\beta}_{CD}^{*\,\prime} W' W \hat{\beta}_{CD}^* \;.$$

3. Verify that SSR_{PC} can be derived as indicated in the footnote on p. 338 of *RAA*.

4. Establish the equivalence between eqn. (10.2.7) and eqn. (10.1.15) when condition (ii) on p. 338 of *RAA* is met.

5. Derive eqn. (10.3.6) on p. 347 of *RAA*.

6. The regression sum of squares for ridge regression is sometimes expressed as $SSR_R = \hat{\beta}_R^{*\,\prime} W' W \hat{\beta}_R^*$, similar to the same expression for least squares (see p. 156 of *RAA*). Is this expression preferable to eqn. (10.3.7) on p. 347 of *RAA*? Why (not)?

7. A "generalized" ridge estimator can be constructed by replacing kI in eqn. (10.3.2) on p. 341 of *RAA* by the matrix $K = \mathrm{diag}(k_1, k_2,\ldots, k_p)$. Show that least squares, principal component, and ridge estimators are special cases of the generalized ridge estimator.

10.2 PROJECTS

10.2.1 Solid Waste Analysis

1. Perform a principal component analysis of the six-variable (SQRTIN and FINAN removed) transformed data set with observations 2, 31, and 40 deleted. Do F tests (or, equivalently, t tests using eqn. (10.1.14) suggest the removal of any components? Which ones (if any)? Compare the ANOVA table for the principal component estimator with the least squares ANOVA table. Are any observed differences in the two tables of sufficient magnitude to conclude that the fit to the responses is substantively different? Explain.

2. Do the latent roots and latent vectors of A A reveal any nonpredictive multi-collinearities? Which ones (if any)?

3. Plot a ridge trace for these predictor variables using, say, 20 values of k in some appropriate interval (e.g. $0 \le k \le .20$). Select a value for k which indicates that the ridge trace has stabilized. Compare the ridge estimates with the least squares estimates and explain any observed differences.

4. Which (if any) of the three biased estimators is preferable to least squares for this data set? Why?

5. After performing any additional analyses you deem appropriate, which set(s) of estimates do you believe is (are) most appropriate for this data set? Why?

10.2.2 Mortality and Air Pollution

1. Using the latent roots and latent vectors of both $X'X$ and $A'A$ assess the predictive value of the multicollinearities which occur in the final subset(s) of predictor variables chosen in Exercise 4, Section 9.2.2. On the basis of this assessment select appropriate principal component and/or latent root estimates for this data set.

2. Select a ridge parameter (a) from the ridge trace and (b) so that the largest VIF is less than

30

10. Compute the corresponding ridge estimates.

3. Discuss the relative merits of all the estimates calculated in Exercises 1 and 2. Choose one set (if possible) as the final estimates for this problem. Carefully state your final conclusions, relative to *this* data set, concerning the association between air pollution and mortality.

APPENDIX A

SMSAs IN DATA SET B.1

Obsn. No.	SMSA	Obsn. No.	SMSA
1.	Akron, OH	31.	Memphis, TN
2.	Albany, NY	32.	Miami, FL
3.	Allentown, PA	33.	Milwaukee, WI
4.	Atlanta, GA	34.	Minneapolis, MN
5.	Baltimore, MD	35.	Nashville, TN
6.	Birmingham, AL	36.	New Haven, CT
7.	Boston, MA	37.	New Orleans, LA
8.	Bridgeport, CT	38.	New York, NY
9.	Buffalo, NY	39.	Philadelphia, PA
10.	Canton, OH	40.	Pittsburgh, PA
11.	Chattanooga, TN	41.	Portland, OR
12.	Chicago, IL	42.	Providence, RI
13.	Cincinnati, OH	43.	Reading, PA
14.	Cleveland, OH	44.	Richmond, VA
15.	Columbus, OH	45.	Rochester, NY
16.	Dallas, TX	46.	St. Louis, MO
17.	Dayton, OH	47.	San Diego, CA
18.	Denver, CO	48.	San Francisco, CA
19.	Detroit, MI	49.	San Jose, CA
20.	Flint, MI	50.	Seattle, WA
21.	Fort Worth, TX	51.	Springfield, MA
22.	Grand Rapids, MI	52.	Syracuse, NY
23.	Greensboro, NC	53.	Toledo, OH
24.	Hartford, CT	54.	Utica, NY
25.	Houston, TX	55.	Washington, DC
26.	Indianapolis, IN	56.	Wichita, KS
27.	Kansas City, MO	57.	Wilmington, DE
28.	Lancaster, PA	58.	Worcester, MA
29.	Los Angeles, CA	59.	York, PA
30.	Louisville, KY	60.	Youngstown, OH

NITROUS OXIDE EMISSIONS DATA

No.	NOX	HUMID	TEMP	BPRES
1	.81	74.92	78.36	29.08
2	.96	44.64	72.48	29.37
3	.96	34.30	75.22	29.28
4	.94	42.36	67.28	29.29
5	.99	10.12	76.45	29.78
6	1.11	13.22	67.07	29.40
7	1.09	17.07	77.79	29.51
8	.77	73.70	77.36	29.14
9	1.01	21.54	67.62	29.50
10	1.03	33.87	77.20	29.36
11	.96	47.85	86.57	29.35
12	1.12	21.89	78.05	29.63
13	1.01	13.14	86.54	29.65
14	1.10	11.09	88.06	29.65
15	.86	78.41	78.11	29.43
16	.85	69.15	76.66	29.28
17	.70	96.50	78.10	29.08
18	.79	108.72	87.93	28.98
19	.95	61.37	68.27	29.34
20	.85	91.26	70.63	29.03
21	.79	96.83	71.02	29.05
22	.77	95.94	76.11	29.04
23	.76	83.61	78.29	28.87
24	.79	75.97	69.35	29.07
25	.77	108.66	75.44	29.00
26	.82	78.59	85.67	29.02
27	1.01	33.85	77.28	29.43
28	.94	49.20	77.33	29.43
29	.86	75.75	86.39	29.06
30	.79	128.81	86.83	28.96
31	.81	82.36	87.12	29.12
32	.87	122.60	86.20	29.15
33	.86	124.69	87.17	29.09
34	.82	120.04	87.54	29.09
35	.91	139.47	87.67	28.99
36	.89	105.44	86.12	29.21
37	.87	90.74	86.96	29.17
38	.85	142.20	87.08	28.99
39	.85	136.52	84.96	29.09
40	.70	138.90	85.91	29.16
41	.82	89.69	86.69	29.15
42	.84	92.59	85.27	29.18
43	.84	147.63	87.25	29.10
44	.85	141.35	86.34	29.06

Reprinted with permission.

APPENDIX B

SOLUTIONS FOR SOLID WASTE ANALYSIS

B.1 CHAPTER 1

1. Inferences cannot be made directly on the total solid waste produced in these regions but only on the estimate from the economic model. Only if additional studies validate the accuracy of the solid waste estimates will inferences be appropriate for solid waste production. Lack of complete information on solid waste produced in the regions necessitated this approach.

2. Inferences could be considered valid if the purpose of an analysis is solely to assess relationships between solid waste production and the land use variables, intentionally ignoring other possibly relevant factors. For example, if a zoning commission desires to use zoning to restrict various types of land use so adequate waste disposal can be insured, an acceptable fit to a data set such as Data Set B.2 would enable the commission to predict the amount of solid waste based on the proposed land use. One must, however, insure that the land use variables provide an acceptable fit to the data and be careful not to imply causality in the inferences.

3. First, only 40 regions in a nine-county area of Southern California were selected. Whether these regions are typical or representative of other areas in California or the U.S. is unknown (from the information given to the reader). Second, the estimates generated by the economic model could further limit the generality of the results since these estimates were based on factors obtained from other research studies, each of which might be less general (more limited) than the present one. Finally, other possibly relevant variables (e.g., population, income, employment) have been excluded from the data base. Even if an acceptable fit to the solid waste variable is achieved with the land use variables, conclusions must be cautiously stated since other predictor variables could produce equally acceptable fits.

4. One instance of extrapolation would occur if prediction equations from this analysis are used to predict solid waste for regions which have land use acreage which exceeds the maximum for any of the variables in Data Set B.2. A generalization would occur if the prediction equations are used for areas of California outside these 40 regions or for regions in other states, even if the land use acreage for these new regions does not fall outside the ranges on the predictor variables in Data Set B.2. An example of a causal inference is if one attempts to conclude from an analysis of the data that one of the predic-

tor variables is the single most important contributor to all solid waste production; e.g., restaurants and hotels always produce more of solid waste than any of the other categories of land use.

B.2 CHAPTER 2

1. Several of the observations are potential outliers based on the raw data values: 2 (WASTE, WHOLE, RETAIL, REST, FINAN, MISC, and HOME have large values); 8 (HOME is large); 10 (small values on each of the predictor variables but not on WASTE); 31 (INDUS is large); 40 (METAL and WHOLE are large). The observations cannot be deleted because (i) there is no theoretical reason nor obvious transcription error which dictate the removal of any observation, and (ii) large or small values on the variables do not necessarily indicate outliers.

2. The following table shows observations which are outside $\bar{X} \pm 2S$ and $\bar{X} \pm 3S$ for each of the individual variables.

EXTREME OBSERVATIONS

Variable	Outside ($\bar{X} \pm 2S$)	Outside ($\bar{X} \pm 3S$)
WASTE	15	2
INDUS	40	31
METAL	35	40
WHOLE		2,40
RETAIL	8	2
REST	8,28	2
FINAN	28	2
MISC	8	2
HOME	28	2,8

Several observations in addition to those noted in the previous exercise are suggested as possible outliers in this table. (Note: Caution should be used with this approach since outliers can greatly alter $\bar{X}$ and inflate S.) Again, this exercise and the previous one should be viewed as only providing a first indication of possible outliers.

3. On the plot of WASTE vs. INDUS points 2 and 31 appear in extreme corners of the plot. It is not a certainty that these points are outliers, however, since the points in the plot fan out from lower left to upper right. Similar comments apply to points 2 and 40 in the plot of WASTE vs. METAL. In the plot of WASTE vs. WHOLE, point 40 is in an extreme corner of the plot. Point 2 appears in the extreme upper right corner of several of the plots, e.g., WASTE vs. WHOLE, RETAIL, REST, FINAN, MISC, HOME. Note the consistency of this last finding with the answers to the previous two exercises: observa-

tion 2 has large values on all these variables.

4. The strongest tendencies in several of the plots is a wedge-shape or fanning-out of the points from the lower left corner of the plot to the upper right. This trend is most obvious in the plots of WASTE vs. INDUS and METAL. In several other plots (e.g., WASTE vs. MISC and HOME) the position of point 15 near the vertical axis suggests a wedge-shape but the removal of that point largely removes the wedge-shape. Since the variability in the wedge-shaped plots increased with the magnitude of the predictor variables, square root or logarithmic transformations are suggested.

5. Points 2, 15, 31, and 40 appear to have the most potential for altering the fit based on their appearance in extreme regions of the above scatterplots. Logarithmic transformations of INDUS and METAL leave a slight upward curvature in scatterplots of WASTE vs. each of these transformed predictor variables. Plots of WASTE vs. SQRTIN (square root of INDUS) and SQRTME (square root of METAL) do not leave an indication of curvature or a wedge-shape, especially if the above four points are ignored. Transformations of the response variable worsen several of the plots so it seems the square root transformation of INDUS and METAL is most appropriate at this stage.

B.3   CHAPTER 3

1. Use of TOTAL precludes the assessment of any association between individual land use variables and WASTE. The "spatial and functional" analysis of solid waste production is thereby restricted.

2. Simplicity and the ability to include other types of predictor variables without having an unmanageable number in the model are the major advantages of combining similar variables into a single composite. One major disadvantage is that if some of the land use variables are ineffective as predictors of WASTE, they can mask the effect of the important ones when all of them are combined in TOTAL.

3. Observations 2 and 15 retain the extreme positions which were obvious on several of the previous plots. There is no apparent fanning or curvature in the plot. Observation 8 appears in an extreme corner of the plot due to its extremely large value of HOME.

4. The coefficient of determination for this fit indicates that less than half ($R^2 = 0.47$) the variability of the response variable is accounted for by TOTAL. The estimated error standard deviation is 0.265, approximately 70% of the average response ($\bar{Y} = 0.380$). Each of these measures indicates that the fit is not sufficiently accurate – too much of the response variability remains unexplained.

5. Proof of the validity of the assumed model is a causal inference. Empirical models are approximations to physical phenomena; even an acceptable fit only confirms the adequacy of this one (of perhaps many) approximation.

6. The removal of observation 2 drops the beta weight from 0.689 to 0.522. Elimination of observation 8 increases it to 0.750. The changes in slope are consistent with the visual elimination of each of these points from the scatterplot; e.g., observation 8 is in the lower

right portion of the scatterplot and its removal should increase the slope.

B.4   CHAPTER 4

1. All the pairwise correlations between pairs of the last 5 predictor variables are approximately 0.8 or larger; in particular, the correlation between RETAIL and MISC is 0.979. There is also a large correlation between METAL and WHOLE, $r = 0.889$.

2. The smallest latent root of $W'W$ is 0.018. In the associated latent vector, the elements corresponding to RETAIL and MISC are -0.644 and 0.737, respectively. Thus, for the standardized predictor variables,

$$-0.644 \, \text{RETAIL} + 0.737 \, \text{MISC} \approx 0$$

    or,
$$\text{RETAIL} \approx \text{MISC}.$$

3. The elements in $\underline{V}_1$ and $\underline{V}_2$ which are 0.4 or larger in magnitude correspond to the last 5 predictor variables, reflecting the multicollinearities among them.

4. The latent vectors $\underline{V}_1$, $\underline{V}_4$ and $\underline{V}_5$ reveal pairwise correlations between RETAIL and MISC, METAL and WHOLE, and FINAN and HOME, respectively. The latent roots associated with these latent vectors are, respectively, 0.018, 0.106, and 0.204. The magnitudes of the latent roots are indicative of the sizes of the respective pairwise correlations (0.979, 0.889, 0.793) and, hence, the strength of the multicollinearities.

5. The strong multicollinearities prevent one from concluding that one of the multicollinear predictor variables is most closely associated with waste production. For example, if one attempted to conclude that retail trade was the most important determinant of solid waste production, the strong pairwise multicollinearity between RETAIL and MISC would cause one to question whether miscellaneous services might be an equally important determinant of solid waste production. As will be shown in Chapter 9 least squares inferences on two variables which are highly multicollinear often do not lead to correct conclusions about the relative importance of the predictor variables.

B.5   CHAPTER 5

1.   ANOVA

Source	d.f.	S.S.	M.S.	F	R^2
Mean	1	5.780	5.780	340.983	
Regression	8	4.570	.571	33.697	.897
Error	31	.526	.017		
Total	40	10.876			

Approximately 90% of the variablility of the response variable is explained by these land use variables. The estimated error standard deviation, $\hat{\sigma} = 0.130$, is approximately 34% of the magnitude of the mean response, $\bar{Y} = 0.380$. Although 90% of the response

37

variability is explained with this prediction equation, the size of $\hat{\sigma}$ indicates that the individual responses are not predicted with a high degree of precision. Additional predictor variables are needed to increase precision.

2.   ANOVA

Source	d.f.	S.S.	M.S.	F	R^2
Mean	1	5.780	5.780	308.595	
Regression	8	4.515	.564	30.128	.886
Error	31	.581	.019		
Total	40	10.876			

The two ANOVA tables are not greatly different. The choice of which fit to prefer should be based on considerations other than these tables.

3. Either of the previous two fits is superior to the single-variable fit of WASTE to TOTAL since, for the latter fit, $R^2 = 0.475$ and $\hat{\sigma} = 0.265$.

4. As the estimates below reveal, comparison of the magnitudes of the raw estimates does not lead to the same conclusion as does comparison of the (unit length) standardized ones. For example, the raw coefficient estimate for SQRTME is approximately two orders of magnitude larger than that for WHOLE but the standardized estimates are much closer to one another.

Predictor Variable	Raw Estimate	Standardized Estimate
SQRTIN	− .00519	-.628
SQRTME	.01153	.693
WHOLE	.00014	.606
RETAIL	-.00036	-.419
REST	.01663	3.045
FINAN	-.00078	-.234
MISC	.00067	.446
HOME	-.00002	-1.348

5. Since each estimated coefficient measures the effect of the adjusted predictor variable on the prediction of WASTE, the coefficient for SQRTIN can be interpreted as measuring the effect of this variable on the response. Since RETAIL is highly multicollinear with other predictor variables, however, its estimated coefficient measures the unique contribution of RETAIL after adjusting for its joint influence with the other predictor variables. Thus, the coefficient for RETAIL does not measure its complete effect on WASTE.

38

B.6 CHAPTER 6

1. In this study land use is not a "controlled variable"; i.e., the researchers could not specify a-priori the distribution of land in a region to the eight categories of land use. Consequently, the land use variables are stochastic and inferences must be conditioned on the observed values of these variables.

2. An examination of the F statistics associated with the mean response (F = 340.983) and the regression effect (F = 33.697) are significant at any reasonable significance level. These statistics test, respectively, that $\alpha^* = 0$ and $\underline{\beta} = \underline{0}$.

3. The (unit length) standardized coefficient for HOME ($\hat{\beta}_8^* = -1.348$) is approximately three times larger than that for MISC ($\hat{\beta}_7^* = 0.446$). The t statistics for testing the significance of the respective coefficients are similarly related with that for HOME (t = -3.395) being about 5 1/2 times larger than that for MISC (t = 0.609). Since the standard errors of any two estimated coefficients generally are not equal in either the raw or standardized form, the t statistics will not necessarily have the same relative magnitudes as either set of coefficient estimates.

4. The following table details the quantities needed for calculation of the confidence intervals.

	$\hat{\beta}_j$	d_j	$\hat{\beta}_j^*$	q_{jj}
SQRTIN	$-.00519$	120.837	-0.628	2.605
MISC	.00067	662.154	0.446	31.692

Confidence intervals for the raw coefficients are
$$-.00874 \le \beta_1 \le -.00165 \text{ and } -.00158 \le \beta_7 \le .00293.$$
For comparison purposes confidence intervals for the standardized coefficients are
$$-1.057 \le \beta_1^* \le -0.199 \text{ and } -1.050 \le \beta_7^* \le 1.942.$$
Observe from the confidence intervals for the standardized estimates that the one for MISC is larger, indicating a greater degree of uncertainty about β_7^* than about β_1^*. This is due to the larger variation in the values of MISC (compare the d_j) and the larger variance for the standardized estimator (recall, $Var(\hat{\beta}_j^*) = q_{jj}\sigma^2$). Note also that the confidence interval for β_7 (or β_7^*) includes zero while that for β_1 (or β_1^*) does not.

5. The general expression for a $100(1-\gamma)\%$ prediction interval when σ^2 is unknown can be written as
$$\hat{Y} - t_{1-\gamma/2}(n-p-1)\hat{\sigma}[1 + n^{-1} + \underline{u}_m'(W'W)^{-1}\underline{u}_m]^{1/2} \le Y \le$$
$$\hat{Y} + t_{1-\gamma/2}(n-p-1)\hat{\sigma}[1 + n^{-1} + \underline{u}_m'(W'W)^{-1}\underline{u}_m]^{1/2}$$
where $\underline{u}_m' = (u_1 - \bar{X}_1, u_2 - \bar{X}_2, ..., u_p - \bar{X}_p)$ contains centered values of the predictor variables at the point where the prediction interval is desired. In this problem prediction is desired at $u_j = \bar{X}_j$, j = 1,...,8. Thus the prediction interval reduces to

$$\hat{Y} - t_{1-\gamma/2}(n-p-1)\hat{\sigma}(1+n^{-1})^{1/2} \le Y \le$$

$$\hat{Y} + t_{1-\gamma/2}(n-p-1)\hat{\sigma}(1+n^{-1})^{1/2}$$

For a 95% prediction interval at $u_j = \bar{X}_j$, $\hat{Y} = \bar{Y} = .380$, $t_{.975}(31) = 2.04$, and $\hat{\sigma} = .130$. Hence a 95% prediction interval for WASTE is

$$0.112 \le Y \le 0.648.$$

The only alteration in the above interval for a confidence interval for $E[Y]$ is that $(1+n^{-1})^{1/2}$ is replaced by $n^{-1/2}$. The resulting 95% confidence interval is

$$0.338 \le E[Y] \le 0.422$$

The difference in the intervals occurs because in the former case when $u_j = \bar{X}_j$ one is estimating $\alpha^* + \epsilon$, while in the latter case one is only estimating α^*.

B.7   CHAPTER 7

1. The partial residual plot for INDUS exhibits a wide scatter of observations for small values of INDUS and a curvilinear trend as INDUS increases. The partial residual plot for METAL clearly suggests the need for a transformation; the plot is decidedly curved. With the exception of perhaps one or two points, the partial residual plots for SQRTIN and SQRTME are much more linear then the previous two.

2. No further transformations are clearly evident.

3. Apart from one or two observations at the upper end of the plot, the bulk of the points appear to approximate a reasonably straight line. Because of the recognized tendency for a few extreme points to deviate from the straight line (see Daniel and Wood, 1971, Appendix 3A), there does not appear to be strong evidence that the errors are non-normal as, for example, was apparent in Figure 7.6.

4. Investigations of scatterplots and summary statistics in the exercises of Section 2.2.1 (above) suggested that observations 2, 15, 31, and 40 were candidate outliers. With the transformed model using SQRTIN and SQRTME the residual statistics for these observations are as follows:

Obsn.	h_{ii}	r_i	t_i	$t_{(-i)}$	D_i
2	.832	.156	2.927	3.384	4.722
15	.753	.091	1.407	1.430	.672
31	.810	.030	.536	.529	.136
40	.885	-.092	-2.090	-2.218	3.738

From an examination of this table it is clear that observations 2 and 40 are having an enormous influence on the estimation of the regression coefficients. Deletion of these two observations and a recompilation of the residual statistics yields a Cook's D_i which is inflated for observation 31: $h_{ii} = .871$, $r_i = -.053$, $t_i = -1.310$, $t_{(-i)} = -1.327$, $D_i = 1.284$. All of the above considerations suggest the deletion of observations 2, 31, and 40. The evidence for observation 15 is less clear: some of the statistics are moderately sized but substantially smaller than those of the above three cases. We choose to retain this observation so the data base will not be unduly restricted by the elimination of too many observations.

5. The beta weights and t statistics for the two fits are shown below. Of the differences shown in the table, those of SQRTIN, MISC, HOME, and perhaps REST appear large. An examination of the scatterplots in Section 2.2.1 (above) indicated that two of the points (2,31) were in extreme corners of the plot for SQRTIN and observation 2 was in an extreme corner of the plots for REST, MISC, and HOME. Comparison of the two sets of scatterplots and partial residual plots suggests that observation 40 has a moderate effect on SQRTME and WHOLE.

Variable	Beta Weight		t Statistic	
	$n = 40$	$n = 37$	$n = 40$	$n = 37$
SQRTIN	-.278	-.029	-2.987	-.197
SQRTME	.307	.354	2.700	3.127
WHOLE	.268	.162	2.283	1.502
RETAIL	-.186	-.424	-.586	-1.468
REST	1.349	1.770	5.631	6.906
FINAN	-.104	-.116	-.518	-.638
MISC	.198	-.405	.609	-1.291
HOME	-.597	-.392	-3.395	-1.778

6. Reevaluation of residual plots, residual statistics, etc. reveal no further abnormalities with regard to outlier or model assumptions.

B.8 CHAPTER 8

1.

Step	Variable Added	F	C_k	Cumulative R^2
1	REST	51.79	42.03	.597
2	MISC	10.95	25.75	.695
3	SQRTME	16.38	8.93	.796
4	HOME	4.56	6.20	.822
5	RETAIL	2.57	5.65	.835
6	WHOLE	2.35	5.43	.847
7	FINAN	.39	7.05	.849
8	SQRTIN	.04	9.00	.849

2. The pairwise correlations of WASTE with the predictor variables for the reduced ($n = 37$)

41

data set are: REST(.772), MISC.(.588), SQRTME(.482), HOME(.512), RETAIL(.564), WHOLE(.532), FINAN(.690), and SQRTIN(.556). Due to the large pairwise correlations among the predictor variables the order of inclusion in forward selection does not match the magnitudes of the pairwise correlations with WASTE.

3. On the basis of the R^2 values shown above, little improvement is noted after HOME is included (step 4). Comparison of the F statistics with F values from Table C.4a (significance level = 0.25) in *RAA* indicates that all the variables through step 6, including WHOLE, should be retained. The C_k values suggest that the first 5 or 6 predictor variables should be included.

4. The table below exhibits the statistics which correspond to the deletion of variables with backward elimination. The R^2 values suggest that a satisfactory fit is obtainable through steps 5 or 6. Using a significance level of 0.25, the F statistics indicate termination of elimination after only two steps. The C_k statistics dictate termination at either step 4 or step 5.

Step	Variable Deleted	F	C_k	R^2
1	SQRTIN	.04	9.00	.949
2	FINAN	.38	7.05	.847
3	MISC	1.73	5.43	.838
4	WHOLE	2.21	5.16	.826
5	HOME	5.81	5.37	.795
6	SQRTME	23.16	9.18	.670
7	RETAIL	13.69	30.34	.579
8	REST	.04	42.03	– –

5. The F statistics for forward selection lead to the deletion of only SQRTIN and FINAN while the C_k values also permit the deletion of WHOLE. SQRTIN and FINAN are again deleted in backward elimination when the F statistic is used as a selection criterion; this subset is acceptable by the R^2 criterion as well. The C_k plot for backward elimination strongly suggests that WHOLE and MISC can also be eliminated. Possible subsets are, therefore, (i) SQRTME, RETAIL, REST, HOME, (ii) these four predictor variables plus WHOLE, and (iii) these five predictor variables plus MISC.

6. The second type of t-directed search requires an examination of the t statistics for all the predictor variables in the transformed (n = 37) data set. These t statistics were computed for Exercise 5, Section 7.2.1. The predictor variables are then sequentially added, in a forward selection fashion according to the magnitude of their t statistics and their calculated C_k values. The following table displays the order of entry of the variables and the corresponding cumulative C_k statistics:

Step	Variable Added	F	Cumulative R^2
1	REST	42.03	.597
2	SQRTME	36.08	.639
3	HOME	11.16	.784
4	WHOLE	12.00	.790
5	RETAIL	5.16	.838
6	MISC	5.43	.847
7	FINAN	7.05	.849
8	SQRTIN	9.00	.849

The smallest C_k value occurs for $k = 6$ ($p = 5$). If the first five predictor variables listed above are included in a basic set and all possible regressions are performed on subsets of the remaining predictor variables the additional C_k statistics are:

k	Variables Added to Basic Set	C_k	R^2
6	NONE	5.16	.838
7	MISC	5.43	.847
7	FINAN	6.72	.840
7	SQRTIN	7.15	.838
8	MISC, FINAN	7.05	.849
8	MISC, SQRTIN	7.42	.847
8	FINAN, SQRTIN	8.68	.840
9	MISC, FINAN, SQRTIN	9.00	.849

This table reveals that there are alternative subsets which might be as useful as the three given in the previous answer.

B.9 CHAPTER 9

1. The variance inflation factors for the reduced ($n = 37$) data set with transformed values for INDUS and METAL are:

Variable:	SQRTIN	SQRTME	WHOLE	RETAIL	REST	FINAN	MISC	HOME
VIF:	4.09	2.38	2.17	15.51	12.22	6.20	18.32	9.03

Predictor variables which appear to be most strongly involved in multicollinearities are RETAIL, REST, MISC, and perhaps HOME.

2. Pairwise correlations indicate a considerable degree of redundancy among the last five predictor variables (see Supplement Exercise 4, Section 4.2.1). The strongest multi-

collinearity ($\ell_1 = 0.036$) involves RETAIL, REST, and MISC, the three predictor variables having the largest variance inflation factors. This relationship is (for standardized variables):

$$0.59 \text{ RETAIL} + 0.36 \text{ REST} \approx 0.69 \text{ MISC}$$

or

$$\text{MISC} \approx 0.8 \text{ RETAIL} + 0.5 \text{ REST}.$$

If this relationship is not spurious, it suggests that the "miscellaneous" category of land use is mostly retail land use and to a lesser extent restaurants and hotels, although a clear-cut delineation of this land use into the two categories identified for that purpose (i.e., RETAIL and REST) might not be possible. On the other hand, this multicollinearity could indeed be spurious or the strong relationship between MISC, RETAIL, and REST could be due to another, unknown source.

3. The multicollinearities do not appear to have an obvious adverse impact on the least squares estimates. The patterns in the signs in $\underline{V}_1$, for example, are not the same as the signs on the coefficient estimates for the corresponding variables.

4. There are several possible alternatives to the ordering of the variables in the forward selection and backward elimination analyses. For example, RETAIL and MISC have a large pairwise correlation ($r = 0.95$). Their correlations with WASTE are very similar (0.56 and 0.59, respectively) and can be considered equal for all practical purposes. The inclusion of MISC at step 2 in forward selection suggests that RETAIL could be an alternative at this step. Other alternatives are possible for several stages of forward selection and backward elimination.

5. The three subsets selected in the last chapter are
   (a) SQRTME, RETAIL, REST, HOME
   (b) SQRTME, RETAIL, REST, HOME, WHOLE
   (c) SQRTME, RETAIL, REST, HOME, WHOLE, MISC.

   All three of the predictor variables that are involved in the strongest multicollinearity (and the four which have the largest variance inflation factors) are included in the third subset listed above. The only one excluded from the first two subsets is MISC. Replacement of one of the variables in the first two sets by MISC does not seem to improve any of the selection criteria. Because of the selection of the third subset by several of the selection criteria in forward selection and backward elimination, we choose to consider it the best one at this stage of the analysis. We recognize the possibility that a different conclusion might be drawn after the analysis using biased estimators.

6. Retracing the analyses performed in Section 2.2.1 and 7.2.1, in particular, leads one to conclude that observations 2, 31, and 40 should still be deleted.

B.10 CHAPTER 10

1. The three smallest latent roots of $W'W$ when SQRTIN and FINAN are eliminated from the data set are 0.039, 0.076, and 0.133. The F statistics for testing the significance of the

individual corresponding principal components are, respectively, 0.81, 9.73, and 66.75. Thus only the first component can be deleted on the basis of these F statistics, even if a very small significance level is chosen. The ANOVA tables for least squares and the principal component estimator with one component deleted are shown below. Note the similarity in the two tables; these similarities remain even if the "components deleted" sum of squares is pooled with the error term.

Least Squares ANOVA

Source	d.f.	S.S.	M.S.	F	R^2
Mean	1	4.018	4.018	341.11	
Regression	6	1.959	.327	27.72	.847
Error	30	.353	.012		
Total	37	6.330			

Principal Component ANOVA

Source	d.f.	S.S.	M.S.	F	R^2
Mean	1	4.018	4.018	341.11	
Regression					
Prin. Comp.	5	1.950	.390	33.10	.843
Comp. Del.	1	.010	.010	.81	
Error	30	.353			
Total	37	6.330			

2. The three smallest latent roots of $A'A$ are 0.027, 0.040, and 0.084. The first elements of the corresponding latent vectors are, respectively, -0.357, 0.127, and 0.112. None of the elements are sufficiently close to zero for one to conclude that they are nonpredictive. Careful examination of the latent vectors also indicates that the multicollinearities identified in the latent vectors corresponding to the two smallest latent roots of $W'W$ are recognized in the latent vectors corresponding to the second and third smallest latent roots of $A'A$. The first elements of these two latent vectors are close to the rough cut-off value of 0.10 suggested in the text, but we choose to retain them because (i) they are not as close to zero as is desirable for a relatively certain conclusion that they are nonpredictive and (ii) in the last chapter an analysis of the coefficient estimates did not lead to a conclusion that the multicollinearities were seriously distorting the least squares estimates.

3. The ridge trace curves slope rather slowly for this data set. It is thus somewhat difficult to select a precise value of k from the ridge trace. (Note: This is consistent with the observations made earlier concerning the lack of drastic effect of the multicollinearities on the least squares estimates.) A small value of k, say 0.03, seems reasonable for this data set. The least squares and ridge regression estimates for standardized predictor variables are shown below. A few moderate changes in the estimates are observable but there are no dramatic alterations of signs or magnitudes.

	Standardized Estimates	
Variable	Least Squares	Ridge (k = 0.03)
SQRTME	.512	.441
WHOLE	.245	.247
RETAIL	− .723	− .540
REST	2.533	2.037
MISC	− .628	− .342
HOME	− .525	− .484

4. Because of the comments made in the last chapter on the effects of the multicollinearities in this data set on the least squares estimates and the above discussions of the latent root and ridge estimates, it appears that biased estimation is not necessary. The elimination of one component by the principal component estimator using F statistics in Exercise 1 could be due to the effect of the smallest latent root on the test statistics (see the discussion on p. 330 of *RAA*); in any case, deletion of a component is not consistent with the indications mentioned above.

5. Conclusion left to the reader.

APPENDIX C

TEXT ERRATA

Chapter	Page	Position	Corrections
2	26	Figure 2.2	Add the point (GNP, LIT) = (2900,30).
	38	line 9 from bottom	$36 \to 32$
	50	Exercise 2, line 2	plots $\to$ points
		Exercise 8, lines 2-3	exp (GNP) $\to$ ln(GNP)
	51	Exercise 9, line 2	Figures 2.2, 2.3, 2.4
3	59	Table 3.3, LIC(1969)	$837.60 \to 837.80$
	83	female suicide rates	$R^2 = \dfrac{54.50}{104.23} = .523$
		line 11	$63\% \to 48\%$
	90	lines 2 and 3	TTS(adj) $\to$ TSS(adj)
	91	Exercise 10, line 2	verify $\to$ compute the quantities shown in
		Exercises 10 and 11	male $\to$ female
4	97	eqn. (4.1.2)	$\displaystyle\sum_{j=1}^{p} \to \sum_{j=1}^{r}$
	99	5 lines after eqn. (4.1.4)	UV is $\to$ VU is
	126	Exercise 2, line 2	eqn. (3.2.4) $\to$ eqn. (3.2.14)
	127	Exercise 7, line 1	length $\to$ unit length
5	134	line 8	$X_{ip} Y_i \to \Sigma X_{ip} Y_i$
	137	Table 5.2, HOM Stage 3	$1.13508 \to -1.13508$
	138	3 occurrences	$\hat{r}_1(Y) \to \hat{r}_{i1}(Y)$
		line 5 from footnote	$\hat{r}_{i1}(X_2) \to r_{i1}(X_2)$
	143	line 14	$\bar{X}_j \beta_j \to \bar{X}_j \hat{\beta}_j$
	145	line 1	$\hat{\beta} \to \hat{\beta}_j$
	146	line 3	$Y^0 \to \underline{Y}^0$
	155	eqn. (5.3.2)	$\hat{\underline{\beta}}'M'\hat{\underline{Y}} \to \hat{\underline{\beta}}'M'\underline{Y}$
	157	Table 5.4	$R \to R^2$
	159	eqn. (5.3.5)	After " = " insert "$\Sigma Y_i^2 - n\bar{Y}^2 - $".
	161	line 13	$\Sigma s_i^2 \to \Sigma s_i^2$
	161-162	2 occurrences	$D_j \underline{Y} \to D_j' \underline{Y}$
		5 occurrences	$D_j D_j \to D_j' D_j$

	162	line 4	$D_j] \underline{Y} \to D_j'] \underline{Y}$
		line 7	$D_j] \underline{X}_j \to D_j'] \underline{X}_j$
	166	Exercise 3, line 3	Section 5.3.2 $\to$ Section 5.2.3
6	180	line 7	$X_1^* \underline{Y} \to X_1^{*\,\prime} \underline{Y}$
	181	last line, 2 occurrences	uncorrelated with $\to$ independent of
	182	line 5 from bottom	Delete "although...uncorrelated."
		line 3 from bottom	Add: If the true model is $Y_i = \alpha + \beta X_i + \epsilon_i$ rather than $Y_i = \alpha + \beta W_i + \epsilon_i$ and X_i and ϵ_i are all mutually independent, the $\beta(u_i - \bar{u})$ term is eliminated from eqn. (6.2.7) and $E[\hat{\beta}] = \beta$. This would occur if the predictors were random variables but measured exactly.
	189	line 16	$u_j \sim N(\mu_j, \sigma^2 C_{jj}^2) \to u \sim N(\mu, \sigma^2 C)$
	193	Table 6.5, t test on α	$X \to \bar{X}$
	194	line 2 from footnote	$\hat{\beta}_j \to \hat{\beta}_j^*$
	207	line 6 from bottom	$Y \to \hat{Y}$
	209	line 2 from bottom	wider $\to$ narrower
	214	line 7	all $t_i, u_i \to$ all t_i, u_j
	215	eqn. (6.B.1)	$AW + \underline{b} \to A\underline{W} + \underline{b}$
		line 9, 3 occurrences	$Y \to W$
	216	line 8	$\underline{0} \to \underline{0}'$
	217	line 2	$\text{Var}[\underline{Y}] \to \Sigma_Y = \text{Var}[\underline{Y}]$
		eqn. (6.C.1)	$\underline{\mu}_Y \to A\underline{\mu}_Y$
	218	Exercise 2, line 1	$\alpha \to \alpha^*$
		Exercise 3, line 6	variances $\to$ variances of the coefficient estimators
7	221	line 15	$(13{,}500 - Y)/Y \to \ln[(13{,}500 - Y)/Y]$
		EXPONENTIAL	$1 + e^{(\ldots)} \to 1 + e^{-(\ldots)}$
	228	line 7	$r_i^2 \to \Sigma r_i^2$
		line 16	Delete "rather...". Add "They can usually be treated as t statistics (individually) with $n - p - 1$ degrees of freedom even though r_i and $\hat{\sigma}$ are not independent."
	231	line 2	Delete "where $\hat{\sigma}^2 = \text{MSE}$."
	248	Figures (7.12) and (7.13)	Interchange figure titles.
	257	line 3 after Table 7.6	$\hat{\underline{B}}_{(-40)} \to \hat{\underline{B}}_{(-19)}$
		line 4 after Table 7.6	$D_{40} = 2.253 \to D_{19} = 2.310$
		line 14 after Table 7.6	Table 7.5 $\to$ Table 7.6
	258	eqn. (7.A.1)	$\underline{B}_{(-i)} \to \hat{\underline{B}}_{(-i)}$
		line 19	The open bracket should precede $(X^{*\prime}X^*)^{-1}$
		last line	$\underline{u}_i^* \to \underline{u}_i^{*\,\prime}$

	259	line 1	$\underline{u}_i \rightarrow \underline{u}_i^*$
8	268	line 8	Replace "is that it..." with "is that if σ^2 is known and inserted for MSE in eqn. (8.1.7), C_k is an unbiased estimator of".
9	295	line 3 from footnote	$\underline{V}_j \rightarrow \underline{V}_r$
	314	Exercise 8	Table 9.4 $\rightarrow$ Table 9.6
10	324	lines 7,9; 4 occurrences	Interchange $\hat{\gamma}_{PC}$ and $\hat{\beta}_{PC}^*$
	331	line 14	$\ell_j \approx 0 \rightarrow \lambda_j \approx 0$
	333	Table 10.8	$\underline{V}_1 \rightarrow \underline{V}_2$
	339	eqn. (10.2.7)	$MSE \rightarrow MSE_{LR}$
App. A	358	PHYS(Australia)	$806 \rightarrow 860$
	360	LIC(1969)	$837.60 \rightarrow 837.80$
	362	Data Set A.5	Numerical results in *RAA* use the original data file which records HUMID and TEMP to two decimal places (see Appendix A, Classroom Supplement).
		BPRES(Obsn. 42)	$28.18 \rightarrow 29.18$
		TEMP (Obsn. 17)	$78.7 \rightarrow 78.1$
App. C	367	Data Set A.8	The numerical results given in *RAA* utilize BRACH and TIBIO measurements which have five decimal points rather than the two reported on p. 367. In order to achieve better accuracy and obtain the same numerical results as appear in the text, compute BRACH and TIBIO as on p. 349 and express them as a percentage with five decimal points.
	375	line 2 from bottom	$H_o: \rho \leq 0$ vs. $H_a: \rho > 0 \rightarrow H_o: \rho \geq 0$ vs. $H_a: \rho < 0$
	387	Table C.5b, footnote	smaller $\rightarrow$ greater
Biblio.	394	Lave and Seskin	$1970 \rightarrow 1979$

Printed and bound by CPI Group (UK) Ltd, Croydon, CR0 4YY

22/10/2024

01777326-0001